.......This page left intentionally blank......

TABLE OF CONTENTS

DISCLAIMER

The information given in the book has been prepared by the Author. The Author requests it be acknowledged as the source of this information. The author believe that all information contained in this book is accurate; however, the user should be aware that the recommendations provided in this book do not replace any standard or regulation.

Although the author have made every effort to ensure that the information in this book was correct at press time, the author do not assume and hereby disclaim any liability to any party for any loss, damage, or disruption caused by errors or omissions, whether such errors or omissions result from negligence, accident, or any other cause.

Author

Kamran Ahmad

ABOUT THIS COURSE

The book has been designed for the information and knowledge of all those who work in H_2S environment mainly in the oilfields. The comprehensive course has seven chapters and covers all aspects of working safely from the hazards and risks of H_2S in oilfields. The course elaborates the properties of H_2S, Its possible presence, its adverse effects on human beings, detection and alarm system for H_2S, personal protective equipment used for H_2S, emergency response, first aid and contingency planning.

Is H₂S a real hazard?

A common hazard that's not easily recognized, but can easily kill, is lurking in crude petroleum and natural gas fields. Occurring naturally at oil and gas sites, hydrogen sulfide (H_2S) is an extremely hazardous gas. High concentrations can cause shock, convulsions, inability to breathe, rapid unconsciousness, coma and death. Effects can occur within as little as a single breath. Exposure to hydrogen sulfide is present for all oil and gas personnel including service companies and contractors. Concentrations are often found to be above the permissible limits, so all personnel must be properly trained and sites must be monitored.

Have you ever known about the adverse effects of H_2S? Or, have you ever come across anyone who has been affected by H_2S while working in the oilfields? Dealing with H_2S is quite a tricky job and requires employees' participation and involvement and management commitment in providing resources for training and acquiring detection equipment. H_2S may turn the oilfields into disaster sites if not dealt professionally.

What is Safety?

Safety is absence of injury and ill health or even the Risk of it.

This definition leads to an ideal state for safety and is quite a challenging job but this is the aim. But how do we manage to achieve the safety? It is managed through a process of Identifying hazards, assessing risk and defining control measures.

This can be replicated to quite a simple example. When you cross a road, you know that accidents happen and you are aware that road crossing is a hazardous activity. Subsequently you assess the risk and you figure out the speed of approaching vehicle and you decide whether to cross the road or let go the approaching vehicle and then cross the road.

Managing H_2S safety is no more different.

Risk Evaluation

Evaluation of risks is known to human since the old ages. If you have to deal with the hazards like naked live electrical conductors, crossing a busy road, working at height or fire then you must be able to manage the associated risk because you are aware of the possible consequences. But,

H₂S is an invisible hazard

Dealing with H_2S in oilfields is most important as the H_2S is the second most poisonous gas known to man and it may kill the first time it is inhaled.

CHAPTER 1
H$_2$S Properties and Characteristics

Introduction

Hydrogen sulfide is a highly toxic, colorless gas that is heavier than air. Because it is heavier than air, it will settle into low lying areas, creating a hazard if left undetected.

"Hydrogen sulfide" refers to either the gaseous or liquid forms of the compound. Synonyms for hydrogen sulfide include hydrosulfuric acid, sulfurated hydrogen, sulfur hydride, rotten-egg gas, and stink damp.

Hydrogen sulfide is formed when two hydrogen atoms bond to an atom of sulfur. The chemical expression for this is H_2S. H_2S is another name for hydrogen sulfide.

Hydrogen sulfide is toxic to humans and most other animals by inhibiting cellular respiration a manner similar to hydrogen cyanide. When it is inhaled or it or its salts are ingested in high amounts, damage to organs occurs rapidly with symptoms ranging from breathing difficulties to convulsions and death. Despite this, the human body produces small amounts of this sulfide and its mineral salts, and uses it as a signaling molecule.

"Occupational exposure to hydrogen sulfide" refers to any workplace situation in which hydrogen sulfide is stored, used, produced, or may be evolved as a consequence of the process.

Hydrogen sulfide is often produced from the microbial breakdown of organic matter in the absence of oxygen, such as in swamps and sewers; this process is commonly known as anaerobic digestion, which is done by sulfate-reducing microorganisms. It also occurs in volcanic gases, natural gas deposits, and sometimes in well-drawn water.

Hydrogen sulfide (H_2S) occurs naturally in crude petroleum, natural gas, volcanic gases, and hot springs. It can also result from bacterial breakdown of organic matter. It is also produced by human and animal wastes. Bacteria found in your mouth and gastrointestinal tract produce hydrogen sulfide from bacteria decomposing materials that contain vegetable or animal proteins.

Hydrogen sulfide is slightly soluble in water and acts as a weak acid. A solution of hydrogen sulfide in water, known as sulfhydric acid or hydro-sulfuric acid, is initially clear but over time turns cloudy. This is due to the slow reaction of hydrogen sulfide with the oxygen dissolved in water, yielding elemental sulfur. Hydrogen sulfide reacts with metal ions to form metal sulfides, which may be considered the salts of hydrogen sulfide. Some ores are sulfides. Metal sulfides often have a dark color.

Hydrogen sulfide is a nearly ubiquitous, acute acting toxic substance. It is a leading cause of sudden death in the workplace. Brief exposures to hydrogen sulfide at high concentrations have caused conjunctivitis and keratitis, and exposures at very high concentrations, have caused unconsciousness, respiratory paralysis, and death.

Hydrogen sulfide is especially dangerous when it occurs in low-lying areas or confined workspaces or when it exists in high concentrations under pressure. As a result, work practices, such as continuous monitoring and the use of specified respiratory protective equipment in certain work situations, are of great importance.

Hydrogen sulfide is a flammable, colorless gas with a characteristic odor of rotten eggs. It is commonly known as hydro-sulfuric acid, sewer gas, and stink damp. People can smell it at low levels.

So key points to remember are:

- Hydrogen sulfide is released primarily as a gas and spreads in the air.
- Hydrogen sulfide remains in the atmosphere for about 18 hours (in cold conditions has been reported to remain for up to 36 hours).
- When released as a gas, it will change into sulfur dioxide and sulfuric acid.
- Since hydrogen sulfide is heavier than air, it can collect in low places.

Besides H_2S, hydrogen sulfide is also referred to as:

- Hydrogen sulphide
- Hydrogene sulfure
- Hydro sulfuric acid
- Sewer gas
- Sour gas
- Stink damp

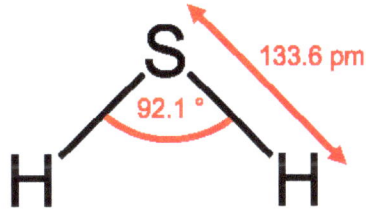

Physical Characteristics

- Color – Clear/Transparent - Colorless/ Invisible
- Odor – It has unpleasant smell like rotten eggs and it has sweetish taste. This is why it is also called rotten egg gas.

At low concentrations, hydrogen sulfide has an unpleasant odor similar to rotten eggs. As concentrations increase, its smell may turn very sweet, but you can never depend on detecting this odor for your safety. The smell of other chemicals may hide H_2S smell. One reason hydrogen sulfide is so dangerous is that it can paralyze your sense of smell. This is called olfactory fatigue and it can occur very rapidly, especially at higher concentrations.

Without proper air monitoring equipment, workers may be unaware they are being exposed to hydrogen sulfide.

Where is H_2S found?

H_2S arises from virtually anywhere where elemental sulfur comes in contact with organic material, especially at high temperatures. Depending on environmental conditions, it is responsible for deterioration of material through the action of some sulfur oxidizing microorganisms. It is called biogenic sulfide corrosion. A portion of global H_2S emissions are due to human activity.

By far the largest industrial source of H_2S is petroleum refineries: The hydrodesulfurization process liberates sulfur from petroleum by the action of hydrogen. The resulting H_2S is converted to elemental sulfur by partial combustion via the Claus process, which is a major source of  elemental sulfur. Other anthropogenic sources of hydrogen sulfide include coke ovens, paper mills (using the Kraft process), tanneries & sewerage.

Hydrogen sulfide is found in nature as a byproduct of decomposing organic matter. It often develops in oxygen depleted environments such as swamps and polluted waters. Hydrogen sulfide can also occur naturally as a component in natural gas, volcanic gases, sulfur deposits etc. This is why protecting workers from hydrogen sulfide is so critical to safe petrochemical and drilling operations. Hydrogen sulfide also exists in many industrial processes, often as a byproduct or waste material.

Vapor Density

Vapor density is defined with respect to air. Air is given a vapor density of one. For this use, air has a molecular weight of 28.97 atomic mass units and all other gas & vapor molecular weights are divided by it to derive their vapor density.

With this definition, the vapor density would indicate whether a gas is denser (greater than one) or less dense (less than one) than air. The density has implications for container storage and personnel safety—if a container can release a dense gas, its vapor could sink and, if flammable, collect until it is at a concentration sufficient for ignition. Even if not flammable, it could collect in the lower floor or level of a confined space and displace air, possibly presenting an asphyxiation hazard to individuals entering the lower part of that space. Unfortunately H_2S has all these properties. H_2S is heavier if compared with air. For equal volumes:

Air = 1

H_2S = 1.189 @ 32 °F.

19% heavier than air.

Flammability range

Before a fire or explosion can occur, three conditions must be met simultaneously. A fuel (i.e., combustible gas) and oxygen (air) must exist in certain proportions, along with an ignition source, such as a spark or flame. The ratio of fuel and oxygen that is required varies with each combustible gas or vapor.

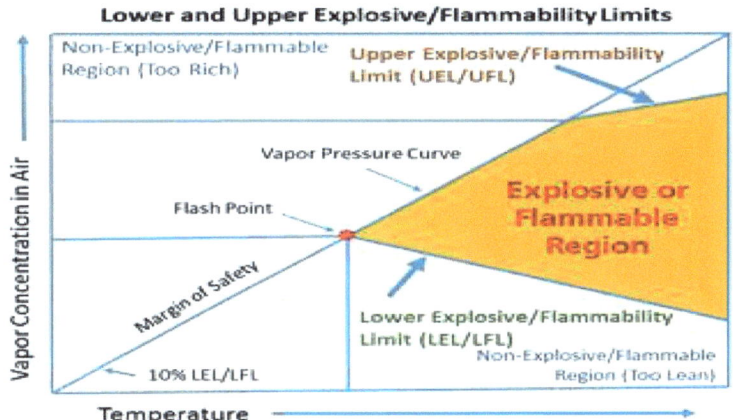

Lower and Upper Explosive/Flammability Limits

The minimum concentration of a particular combustible gas or vapor necessary to support its combustion in air is defined as the Lower Explosive Limit (LEL) for that gas. Below this level, the mixture is too "lean" to burn. The maximum concentration of a gas or vapor that will burn in air is defined as the Upper Explosive Limit (UEL). Above this level, the mixture is too "rich" to burn. The range between the LEL and UEL is known as the flammable range for that gas or vapor. H_2s is highly flammable gas. Its burn range is from 4.3% to 46%. I.e., 43,000 ppm to 460,000 ppm.

How H_2S can deceive?

H_2s has the ability to concentrate, especially when becoming soluble with production waters etc. It is not equally concentrated when soluble. H_2s is heavier than air but is not always at ground level. It can accumulate against an obstruction like building, tree etc.

By-products of Burning

When H_2S is burned, it produces sulfur dioxide. Short-term exposures to high levels of sulfur dioxide can be life-threatening. Exposure to 100 ppm of sulfur dioxide is considered immediately dangerous to life and health (IDLH).

Sulfur dioxide may cause heart problems and respiratory disorders in younger children and elders.

Auto Ignition Temperature

The auto ignition temperature or kindling point of a substance is the lowest temperature at which it will spontaneously ignite in a normal atmosphere without an external source of ignition, such as a flame or spark. Hydrogen sulfide will automatically ignite at 500 ^{0}F. For the reference, diesel exhaust has a temperature in the range of 600-2400 ^{0}F.

H_2S Effects on Metals

Hydrogen sulfide corrosion and its prevention is an important topic in a range of industrial processes and environments including oil and gas and its related activities. H_2S may react with iron and steel causing hydrogen embrittlement and sulfide stress cracking. This will lower the strength of metals and can be disastrous in case of pressure vessels. That's why it is recommended that ventilation systems shall be inspected for corrosion, subjected to regular preventive maintenance, and cleaned at least every 6 months to ensure effectiveness, which shall be verified by periodic airflow measurement at least annually or more frequently according to the judgment of an industrial hygienist. Tempered makeup air shall be provided as required to workrooms in which exhaust ventilation is operating.

Lines and fittings which may carry hydrogen sulfide shall be made of appropriate materials and must be inspected frequently for corrosion, embrittlement, and leaks.

Also recommended in a workplace is that all hydrogen sulfide equipment, including valves, fittings, and connections, shall be checked for tightness and good working order. Such inspections shall be made immediately after new connections are made and after hydrogen sulfide is introduced. Needed repairs and adjustments shall be made promptly.

Because hydrogen sulfide may readily cause pipes and valves to corrode or become brittle, lines and fittings likely to contain hydrogen sulfide should be inspected frequently and receive special attention, monitoring, and maintenance to prevent leaks. Besides the primary work force, the support, maintenance, and repair personnel should be trained in the dangers of hydrogen sulfide, the meaning of alarms, and evacuation procedures.

Chapter Notes:

CHAPTER 2
Health Effects of H$_2$S

Health Hazards of H₂S Gas

Most of the information on human health effects from hydrogen sulfide exposure comes from accidental and industrial exposures to high levels. Exposure to high levels can cause muscle cramps, low blood pressure, slow respiration and loss of consciousness. Brief exposures to high concentrations of hydrogen sulfide (greater than 500 ppm) can cause a loss of consciousness and possibly death. Short-term exposure to moderate amounts of hydrogen sulfide in the workplace produces eye, nose and throat irritation, nausea, dizziness, breathing difficulties, headaches and loss of appetite and sleep. Continued exposure can irritate the respiratory passages and can lead to a buildup of fluid in the lungs.

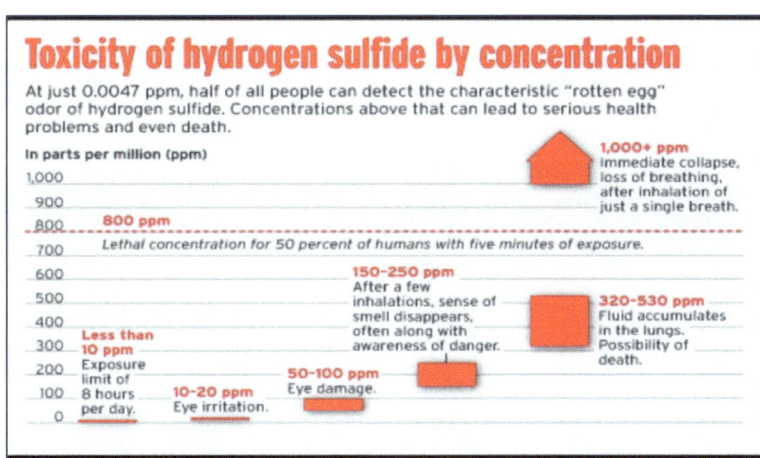

What is PPM or Parts Per Million?

It is unit of measurement of gases in the air. 1PPM of H₂S means 1 particle of H₂S in 1,000,000 particles of Air.

1 / 1,000,000= 1PPM.

Hydrogen sulfide can be measured in exhaled air, but samples must be taken within 2 hours after exposure to be useful. A more reliable test to determine if you have been exposed to hydrogen sulfide is the measurement of thiosulfate levels in urine. This test must be done within 12 hours of exposure. Both tests require special equipment, which is not routinely available in a doctor's office. Samples can be sent to a special laboratory for the tests. These tests can tell whether you have been exposed to hydrogen sulfide, but they cannot determine exactly how much hydrogen sulfide you have been exposed to or whether harmful effects will occur.

The most dangerous aspect of hydrogen sulfide results from olfactory accommodation and/or olfactory paralysis. This means that the individual can accommodate to the odor and is not able to detect the presence of the chemical after a short period of time. Olfactory paralysis occurs in workers who are exposed to 150 ppm or greater. This occurs rapidly, leaving the worker defenseless. Unconsciousness and death has been recorded following prolonged exposure at 50 ppm.

0.02 ppm	No Odor
0.13 ppm	Minimal Perceptible Odor
0.77 ppm	Faint, but Readily Detectable Odor
4.60 ppm	Easily Detectable Odor, Moderate Odor
27.0 ppm	Strong, Unpleasant Odor, but Not Intolerable

The effects on an individual differ greatly depending on factors such as age, physique, measures taken, environment etc. The following data can be used as a guideline when working in an H_2S environment.

- 1-5 ppm- moderately offensive odor, possibly with nausea, or headaches with prolonged exposure.

- 20-50 ppm- nose, throat and lung irritation, digestive upset and loss of appetite.

- 100 -200 ppm- severe nose, throat and lung irritation, ability to smell odor completely disappears.

- 250-500 ppm- potentially fatal build-up of fluid in the lungs (pulmonary edema) in the absence of central nervous system effects (headache, nausea, dizziness), especially if exposure is prolonged.

- 500ppm- severe lung irritation, excitement, headache, dizziness, staggering, sudden collapse ("knockdown"), unconsciousness and death within 4-8 hours, loss of memory for period of exposure.

- 500-1000 ppm- respiratory paralysis, irregular heartbeat, collapse, and death. The symptoms of pulmonary edema, such as chest pain and shortness of breath, can be delayed for up to 48 hours after exposure.

- ▶ **Permissible Exposure Limits (PEL): 10 ppm / 8hr. TWA (Time Weighted Average).**
- ▶ **Short Term Exposure Limit (STEL): 15 ppm / 15 min.**
- ▶ **Immediately Dangerous to Life or Health (IDLH): 100 ppm**

Routes of Entry:

H_2S may enter human body through following routes:-

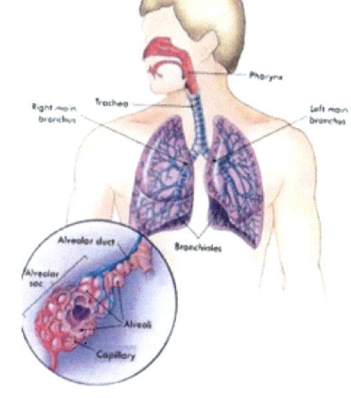

- Inhalation: It is the main source of H_2S entry into human body. H_2S paralysis the olfactory nerves and thus enters the human lungs and nervous system. Workers are primarily exposed to hydrogen sulfide through inhalation. The effects depend on how much hydrogen sulfide you breathe and for how long. Exposure to very high concentrations can quickly lead to death.

- Injection: There is rarely a possibility of H_2S injected into human body.

- Ingestion: Hydrogen sulfide is a gas, so you would not likely be exposed to it by ingestion.

- Absorption: It can enter your body through the skin. When hydrogen sulfide comes into contact with skin, it is absorbed into the blood stream & distributed throughout the body. But much smaller amounts can enter your body through the skin.

Exposure to hydrogen sulfide occurs primarily by inhalation but can also occur by ingestion (contaminated food) and skin (water and air). Once taken into the body, it is rapidly distributed to various organs, including the central nervous system, lungs, liver, muscle, etc.

Target Organs

Hydrogen Sulfide is a toxic irritant gas whose major effects are exerted on the nervous system, the eyes and the respiratory tract. Other target organs affected are the brain and olfactory nerves.

- Olfactory Nerves: Loss of sense of smell occurs within 3-5 minutes at 100 ppm.

- Brain: Headaches, nausea, dizziness, confusion, brain damage.

- Eye irritation: Tearing, inflammation, conjunctivitis, temporary loss of vision.

- Respiratory tract throat irritation: Coughing, olfactory fatigue, pulmonary edema, respiratory arrest.

- Nervous system: H_2S in the bloodstream reduces the oxygen-carrying capability of the blood which depresses the nervous system. Paralysis and respiratory arrest is usually followed within 5-10 minutes followed by cardiac arrest.

Acute & Chronic Exposure Effects

ACUTE: Short term, generally defined as exposure to high concentrations for short durations.

CHRONIC: Long term, generally defined as exposure to low concentrations for a longer duration.

Effects of Acute Exposure

- Olfactory paralysis: The most dangerous aspect of hydrogen sulfide results from olfactory accommodation and/or olfactory paralysis. This means that the individual can accommodate to the odor and is not able to detect the presence of the chemical after a short period of time. Olfactory paralysis occurs in workers who are exposed to 50 ppm or greater. This occurs rapidly, leaving the worker defenseless. Unconsciousness and death has been recorded following prolonged exposure at 50 ppm.

- Eye irritation: Exposure to H_2S will cause eye irritation. This is because eyes are wet and H_2S reacts with water to form a weak acid hydro sulfuric acid or sulfhydric acid. Headache Severe headache may result from short term high concentration exposure of H_2S. Nausea it is a feeling of sickness with an inclination to vomit.

- Headaches: The victim of long time low concentration exposure of H_2S will suffer frequent headaches.

- Nausea: Feeling of sickness with an inclination to vomit will develop.

- Dizziness: It is not a disease but a symptom of other disorders. Vertigo and disequilibrium may cause a feeling of dizziness, but those two terms describe different symptoms. Vertigo is characterized by a feeling of spinning. Disequilibrium is a loss of balance or equilibrium.

- Staggering gait: Due to Nausea and Dizziness, the victim of H_2S may not walk normally and style of walking will be changed.

- Coughing: H_2S affects lungs so coughing will result from exposure.

- Pulmonary edema: It is fluid accumulation in the lungs, which collects in air sacs. This fluid collects in air sacs in the lungs, making it difficult to breathe. It leads to impaired oxygen/CO_2 exchange and may cause respiratory failure.

- Respiratory arrest and brain damage: It prevents delivery of oxygen to the body. Lack of oxygen to the brain causes loss of consciousness. Brain injury is likely if respiratory arrest goes untreated for more than three minutes, and death is almost certain if left untreated for more than five minutes.

- Cardiac arrest: Sudden, sometimes temporary, cessation of the heart's functioning.

Physiological Responses to Acute Exposures

Physiological responses to acute exposure to hydrogen sulfide have been reported as follows:

10 ppm	Beginning of Eye Irritation
50-99 ppm	Slight conjunctivitis and respiratory tract irritation after one hour
100 ppm	Coughing, eye irritation, loss of sense of smell after 2-15 minutes. Altered respiration, pain in the eyes, & drowsiness after 15-30 minutes Throat irritation after an hour. After several hours increase in severity of symptoms Death within 48 hours of exposure
200-300 ppm	Marked conjunctivitis and respiratory tract irritation after one hour
500-700 ppm	Loss of consciousness and possibly death in 30 minutes to one hour
700-1,000 ppm	Rapid unconsciousness, cessation of respiration, and death
1,000-2,000 ppm	Unconsciousness, cessation of respiration an death in a few minutes

Effects of Chronic Exposure

- Eye irritation: As explained earlier, Exposure to H_2S will result into eye irritation as the eye contains water. H_2S reacts with water to form a weak acid hydro sulfuric acid or sulfhydric acid.

- Corneal blistering: The cornea is the transparent tissue that covers the front of the eye. Tiny blisters will form on the corneal surface. When these blisters burst, they are extremely painful.

- Irritation of respiratory tract: Victim will develop the irritation in the respiratory tract leading to coughing and difficulty in breathing.

- Pulmonary edema: Is fluid accumulation in the lungs, which collects in air sacs. This fluid collects in air sacs in the lungs, making it difficult to breathe. It leads to impaired oxygen/CO_2 exchange and may cause respiratory failure.

- Anorexia: It is lack or loss of appetite for food.

- Sleep disturbances: Chronic exposure will result into sleep disorder in the victim.

- Headaches: The victim of long time low concentration exposure of H_2S will suffer frequent headaches

- Nausea: Feeling of sickness with an inclination to vomit will develop.

Who are easy target of H_2S?

- Individuals with eye or respiratory tract problems are especially vulnerable.

- Individuals with anemia (a condition in which there is a deficiency of red cells or of hemoglobin in the blood decreasing the oxygen carrying capability in the blood).

- Alcoholics or those that have consumed alcohol within 24 hrs. of exposure.

- Persons with psychiatric problems.

Factor effecting H_2S exposure

Following factors will determine the effects of exposure:

1. Time: how long has one been exposed to H_2S?
2. Frequency: how many times (over a specified period) has one been exposed to h_2s?
3. Concentration: how high was the concentration of H2S when the victim was exposed?

4. Variables associated with the individual: same exposure may kill one person but the other may survive depending upon the following factors like age, physical strength / health, smoking, etc.

Exposure Limits

The exposure limits to Hydrogen Sulfide are governed by various government regulatory agencies. The following are the different exposure limits recommended by different agencies.

- OSHA General Industry PEL (Permissible Exposure Level - 29 CFR 1910.1000 Table Z-2). 20 ppm ceiling for 10 minutes once, only if no other measurable exposure occurs; 50 ppm peak.

- OSHA Construction Industry PEL (29 CFR 1926.55 Appendix A): 10 ppm (or 15 mg/m3) TWA

- ACGIH: 10 ppm (14 mg/m3) TWA; 15 ppm, 21 mg/m3 STEL (Short Term Exposure Level)

- NIOSH REL (Recommend Exposure Level): 10 ppm Ceiling for 10 minutes.

- ACGIH: 300 ppm is considered by the as immediately Dangerous to Life and Health (IDLH).

Ceiling Concentration 50 ppm / once 10 min.

Human Lethal Concentration 100 - 800 ppm / 5min.

Chapter Notes:

CHAPTER 3
AIR MONITORING & DETECTION OF H₂S

How to Detect H₂S?

Primarily there are three methods of detection of H_2S namely:-

1. Detection through smell by human nose.

This method is highly not recommended and is completely unreliable.

Limitations

- Olfactory Accommodation/Paralysis Occurs About 50-100 ppm.

- Smell of other chemicals may hide the smell of H_2S.

- It cannot distinguish the hazardous concentration of H_2S as nose detects H_2S as low as 0.021 PPM.

- Nose cannot be relied as it is a matter of life and death.

2. Detection through Chemical Reaction:

Lead (ii) Acetate Paper is used to detect H_2S. Its color changes to grey.

Limitations

- May not feasible to be used in most of the workplace environments.

- Requires expertise of the person examining the Detection.

- Does not generate an alarm and requires to be monitored continuously.

3. Electronic Detection Systems:

These systems utilizes Electro-Chemical sensor to detect the presence of H_2S in the environment.

These Detection systems are available in following type:-

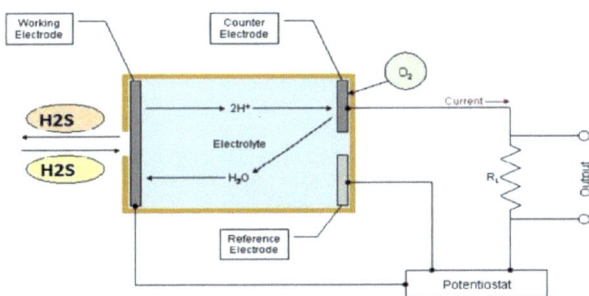

Personal H₂S Monitors: These are hand held personal monitors which provide the reading of H_2S in digital format. Portable hydrogen sulfide monitors should be used to supplement or replace fixed monitors when air currents may move released hydrogen sulfide away from a fixed detector or when the area of hydrogen sulfide release cannot be predicted. A portable hydrogen sulfide detector should have an alarm set to trigger at a hydrogen sulfide concentration of 50 ppm or lower. A two-stage alarm may be desirable on portable monitors.

Advantages

- Portable.
- Cheap solution.
- Workers may use with short training.

Limitations of Personal H₂S Monitors

- May not be feasible to be used in most of the workplace environments.
- Requires expertise of the person examining the Detection.
- May not be used to monitor TWA.
- Can only be used to detect the presence at that point in time when used.
- Requires frequent calibration.
- Require bump testing before use.

Bump Testing is a method of ascertaining the serviceability of the H₂S monitors. This method is used to know that the detection system is responding when exposed to H₂S gas. In this method, hose is kinked and pressure from cylinder is released which fills out the balloon. The hose is then straightened and personal monitor is suddenly exposed with the gas held in the balloon.

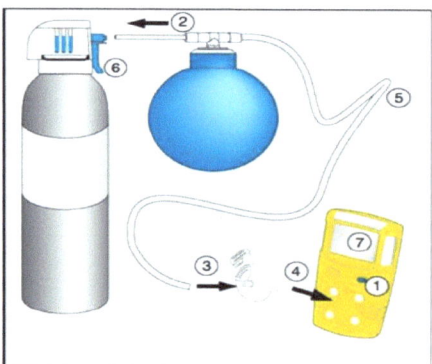

Fixed H₂S Monitors: These are fixed monitors installed over various locations where H₂S might be present. These are equipped with audible Alarm and beacon light which operates over a set value.

Fixed hydrogen sulfide detector systems must have a two-stage, spark proof alarm: the lower triggering level must be set no higher than 10 ppm of hydrogen sulfide to warn workers that hydrogen sulfide is present above the ceiling limit, and the higher triggering level set no higher than 50 ppm of hydrogen sulfide to signal workers to evacuate the area and to obtain respiratory protection for rescue or repair efforts or for carrying out contingency plans.

Advantages

- Reliable.

- Does not require worker competency.

- Can be used for data logging.

Limitations of Fixed H$_2$S Detection systems

- Requires periodic calibrations.

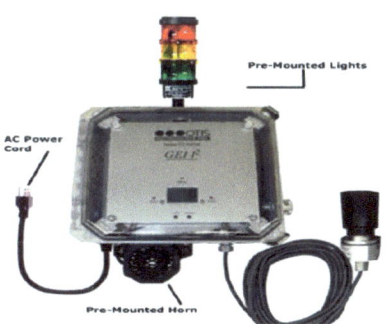

Because hydrogen sulfide can cause fatigue of the sense of smell or actual anosmia, a worker can enter an area where high concentrations of hydrogen sulfide are present without knowing it. There should be constant monitoring with an automatic audible (or audiovisual) warning device in places where a sudden release of hydrogen sulfide might not be expected and would otherwise not be recognized. Fixed continuous monitors may be used to activate ventilation or shut down processes at preset hydrogen sulfide concentrations.

Monitoring Devices Efficiency

Following factor may determine the efficiency of a H$_2$S Detection device:

- High or Low Temperatures: At extreme high or low temperatures, the devices may give erroneous readings and may generate false alarms. Moreover, the device may not generate the alarm even if the H$_2$S concentration value in the environment is reached. Please read the necessary instruction of use on the product.

- Humid or dry weather: Humid or Dry weather may result into erroneous readings of detection systems.

- Frequent Calibration: It is the requirement that Detection Systems must be calibrated in accordance with (Original Equipment Manufacturer) OEM's guidelines at the given intervals. Failing to this, the device may not give the accurate results.

Monitoring Devices Selection

Following factors must be considered while selecting an H_2S monitoring device:-

- **Environmental conditions:** The efficiency and performance of the detection systems highly rely on the environmental conditions. Extreme hot or cold temperatures, dry or humid environment, dust, sand may play important role. So, before the selection of suitable detection system, it is imperative to consider the environment in which the systems are to be used.

- **Spark Proof Casing:** H_2S is a flammable gas. Therefore, it is necessary that all detection systems must be spark/ explosion proof to avoid the risk of fire/ explosion.

- **Detection Range for H_2S:** What detection range is required? It is normally represented in PPM. It is important that the detection system must measure and show the minimum level of H_2S which is hazardous to human.

- **Resolution of Display:** What resolution of the display is required for the specific nature of job it will be used. It is recommended that maximum resolution should not be more than 1PPM if not less.

- **Response time for sensors:** The electro-chemical sensors have certain response time. It is the time once the H_2S comes in contact with the sensor & the sensor processes the value and generates the alarm. Lower the response time, higher will be its effectiveness for saving human lives from ill health.

- **Compatibility with Alarm System:** The detection system should be compatible to a wide range of alarm systems and beacon lights. A detection system is only effective if it serves its intended purpose i.e. to alarm the workers for the presence of H_2S.

- **Specific Training requirement for operation & Maintenance:** Does the detection system require the specific training of all or some of the employees for its day to day operations and maintenance? If it is so, does the organization have sufficient trained and experienced personnel available who may get the training?

- **Calibration frequency requirement:** What is the calibration frequency? How many times of month/ year, the detection system will require to be calibrated. Does the organization have calibration facility and skill available?

- **Accuracy, Reliability & Repeatability:** How accurate are the results of the detection system. How reliable is the detection system for a prolonged use. And what is the repeatability error (If any).

- **Rugged and temper proof:** Is it rugged, robust and tamper proof as per the environmental conditions and cannot be tampered by conscious or unconscious acts of workers performing duties in the area.

Do's and Don'ts of Electronic Detection system

- Installation, calibration and maintenance by authorized and trained personnel only.

- Keep a record of the maintenance and calibration.

- Frequently check the serviceability of the detection and alarm system even if there has been no H2S presence for a longer period of time.

- Do not rely on a single detection system in a large area.

- Install the fixed detection system in the area where high presence of H_2S may be obvious.

- If you smell H_2S and detection system does not alarm, immediately follow company specific evacuation and rescue procedures.

- If damage to the detection or alarm system is seen, immediately replace the system even if it has been calibrated the same day.

- When the detection and alarm systems are being replaced, or being sent for calibration, make sure to continuously monitor H_2S presence in the specific area during this time frame.

- Keep an inventory of the parts/spares and complete sets of detection system to replace the unserviceable system.

- Make sure that alarm is audible to the farthest person in the area in the maximum noise conditions.

- Employees must know the location of fixed detection systems and they must be trained how to respond and not to tamper the system what so ever is the case. In case the employee observes any broken wire, loose connection, damage to equipment, or frequent changing readings of the monitoring system, they must know whom to inform and accordingly inform the concerned person. A record of all such incidents may be maintained

Chapter Notes:

CHAPTER 4
Control Measures for H$_2$S Gas

ENGINEERING CONTROLS

While engineering controls should be used to keep airborne hydrogen sulfide below the concentrations at which it is hazardous to the health of workers, certain situations, such as vessel entry, non-routine maintenance or repair operations, or emergencies, may require respiratory protection. The respirators should be immediately accessible to employees in emergency situations.

The following types of engineering controls are available.

Ventilation: Although ventilation may be present through natural source i.e. wind but it may not be sufficient to completely disperse the H_2S from the work area. That is the reason, forced ventilation system are used.

They are of the following types:- ˙

1. **Positive Pressure Ventilation:** This is similar to a fan which pushes the air away from the work area.

2. **Negative Pressure Ventilation:** This is similar to a vacuum cleaner. It sucks the air/ H_2S from the source and releases it to some far off area.

Use a local exhaust ventilation and enclosure, if necessary, to control amount in the air. It may be necessary to use stringent control measures such as process enclosure to prevent product release into the workplace.

Use non-sparking ventilation systems, approved explosion-proof equipment and intrinsically safe electrical systems in areas where this product is used and stored.

Use a ventilation system separate from other exhaust ventilation systems. Filter the contaminated air before it is directly exhausted to the outside. Use leak and fire detection equipment and an automatic fire suppression system.

Flare Stack: Gas flare stacks are used to burn off excess flammable gas in petrochemical extraction/refining operations, with the classic example of flaring extracted gas at an oil well to relieve an overpressure condition. This open air burning is under heavy scrutiny due to the high volume of emissions. H_2S present in the gas

entering the flare will largely combust to SO_2, a heavily regulated pollutant and contributor of acid rain. H_2S levels in the flare and SO_2 emissions are regulated which forbids burning any fuel gas containing H_2S in excess of 230 mg/dscm. In order to heed these restrictions and their counterparts in jurisdictions around the world, operators require a system for continuous monitoring of H_2S loading in their flare gas. The H_2S Analyzer is a proven solution for online H_2S measurement in demanding applications.

Venting: Venting is similar to ventilation but there is no equipment used to create positive or negative pressure to disperse H_2S. Instead it's like opening the windows and fresh air will come through.

Dispersion of H_2S

- A dispersion plan must be obtained on all drilling locations.

- Public proximity will be addressed and evacuation needs addressed.

- Zones of possible exposure within the area will be addressed as per the concentrations.

- Topography will be a factor.

Factors contributing to poor dispersion

- Slow moving winds (below 10mph).

- High H_2S concentration.

- Low level surfaces.

- Buildings and other obstructions.

- Trees and bushes

- Factors contributing to good dispersion.

- Winds speed above 10 mph.

- Low H_2S concentrations.

- Flat ground.

- No buildings or obstructions.

Factors contributing to good dispersion

- Winds speed above 10 mph.

- Low H_2S concentrations.

- Flat ground.

- No buildings or obstructions.

Other Methods of Control

- **Safety Meetings:** Safety meetings with focus on H_2S Awareness and training for all personnel working in the oilfields.

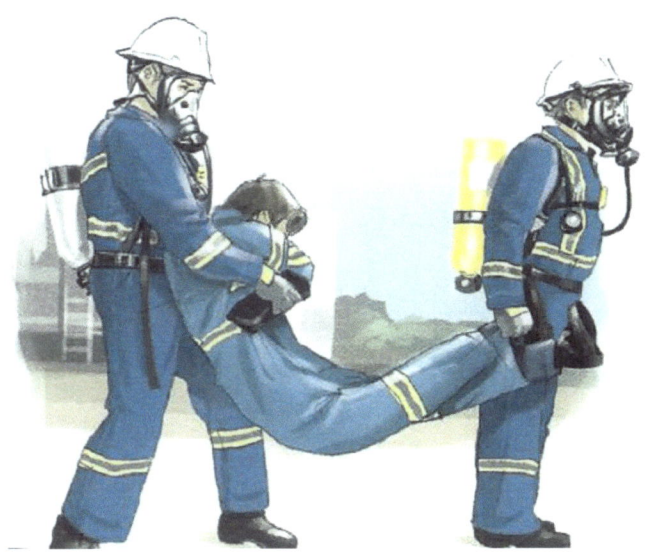

- **Education, Awareness and Training:** The workers must know the adverse consequences of exposure to H_2S. They must be trained how to respond to an emergency situation and how to use PPE's.

- **Buddy System:** A co-operative arrangement whereby individuals are paired or teamed up and assume responsibility for one another's welfare or safety.

- **Eliminate Ignition Sources**: H_2S may explode at 43,000 PPM without any external ignition source. It will burn even at lower PPM level if external ignition source is provided.

- **Keep non-essential personnel out of area:** Only essential people should be present in the work area. Non-essential people must be restricted access to possible H_2S areas.

- **Checking Safety Equipment:** Routine maintenance and calibration of safety devices including detection and alarm system.

- **PPE Maintenance and testing:** PPE's should be well maintained and kept in serviceable condition all the time.

ADMINISTRATIVE CONTROLS

Training Requirements for Workers Who Have a Potential for Exposure to Hydrogen Sulfide:

- Identification of the characteristics, sources, and hazards of Hydrogen Sulfide.

- Proper use of the Hydrogen Sulfide detection methods used on the site.

- Recognition of, and proper response to, Hydrogen Sulfide warnings at the workplace.

- Symptoms of Hydrogen Sulfide exposure.

- Proper rescue techniques and first-aid procedures to be used in a Hydrogen Sulfide exposure.

- Proper use and maintenance of personal protective equipment. Demonstrated proficiency in using P.P.E should be required.

- Worker awareness and understanding of work-place practices and maintenance procedures to protect personnel from exposure to Hydrogen Sulfide.

- Wind direction awareness and routes of egress.

- Confined space and enclosed facility entry procedures.

- Locations and use of safety equipment.

- Locations of safe briefing areas.

- Use and operation of all Hydrogen Sulfide monitoring systems.

- Emergency response procedures, corrective action, and shutdown procedures.

- Effects of Hydrogen Sulfide on the components of the Hydrogen Sulfide handling system.

- The importance of drilling fluid treating plans prior to encountering Hydrogen Sulfide.

All well-drilling sites, should be classified according to areas of potential and/or actual exposure to Hydrogen Sulfide. The recommendations and employee instruction will vary depending on the type of area.

WARNING
HAZARDOUS
AREA
HYDROGEN SULFIDE

EXTREME HEALTH
HAZARD
FATAL OR HARMFUL
IF INHALED

No Hazard Condition

Any well that will not penetrate a known Hydrogen Sulfide formation would be categorized as a No Hazard Area.

API Condition I – Low Hazard

Work locations where atmospheric concentrations of Hydrogen Sulfide are less than 10 ppm.

Safety Recommendations:

- Hydrogen Sulfide warning signs with green flag warning device present.
- Keep all safety equipment in adequate working order.
- Store the equipment in accessible areas.

API Condition II – Medium Hazard

Work locations where atmospheric concentrations of Hydrogen Sulfide are greater than 10 ppm and less than 30 ppm.

Safety Recommendations:

- Legible Hydrogen Sulfide warning sign with yellow flag warning device present.
- Keep a safe distance from dangerous locations if not working to decrease danger.
- Pay attention to audible and visual alarm system.
- Follow the guidance of the operator representative.
- Keep all safety equipment in adequate working order.
- Store the equipment in accessible locations.
- An oxygen resuscitator
- Calibrated, metered Hydrogen Sulfide detection instrument.

API Condition III – High Hazard

Work locations where atmospheric concentrations of Hydrogen Sulfide are greater than 30 ppm.

Safety Recommendations:

- Post legible Hydrogen Sulfide warning signs with red flag warning device.

- Post signs 500 feet from the location on each road leading to the location, warning of the Hydrogen Sulfide hazard.

- Check all Hydrogen Sulfide safety equipment to ensure readiness before each tour change.

- Establish a means of communication or instruction for emergency procedures and maintain them on location, along with contact information of persons to be informed in case of emergencies.

- Ensure usability of two exits at each location.

- Do not permit employees on location without Hydrogen Sulfide safety training.

- Pay attention to audible and visual alarm systems.

- Store the equipment in accessible locations.

- Two Hydrogen Sulfide detectors should be present (one metered the other pump type with detector tube).

- Oxygen resuscitator

- Three wind socks and streamers.

- Two NIOSH/MSHA 30-minute, self-contained breathing apparatus for emergency escape from the contaminated area only.

Hydrogen Sulfide gas is very corrosive and causes metals to become brittle. Therefore, employers need to take special precautions when choosing equipment when they may reasonably expect to encounter Hydrogen Sulfide. This may include appropriate Hydrogen Sulfide trimming of equipment in accordance with the National Association of Corrosion Engineers and/or International Standards Organization (ISO 15156, Petroleum and natural gas industries – Materials for use in Hydrogen Sulfide contagion environments in oil and gas production).

Physical Data for Hydrogen Sulfide

- CAS Number: 7783-06-4

- Synonyms: Sufureted hydrogen, hydrosulfuric acid, dihydrogen sulfide

- Chemical Family: Inorganic sulfide

- Chemical Formula: H_2S

- Normal Physical State: Colorless gas, slightly heavier than air.

- Vapor Density at 59° F and 1 atmosphere: 1.189 (specific gravity)

- Auto Ignition Temperature: 500° F

- Boiling Point: -76° F

- Flammability Limits: 4.3 – 46 percent vapor by volume in air

- Solubility: Soluble in water and oil

- Combustibility: burns with a blue flame to produce sulfur dioxide (SO_2)

PERSONAL PROTECTIVE EQUIPMENTS (PPE)

PPE's for H$_2$S

Employers shall use engineering controls and safe work practices to keep exposure to hydrogen sulfide below the prescribed limits. When necessary, these shall be supplemented by the use of personal protective equipment.

Emergency equipment shall be located at clearly identified stations within the work area and shall be adequate to permit all employees to escape safely from the area. Protective equipment suitable for emergency use shall be located at clearly identified stations outside the exposure area.

Eye/Face Protection: Wear chemical safety goggles. A face shield (with safety goggles) may also be necessary.

Skin Protection: Wear chemical protective clothing e.g. gloves, aprons, boots. In some operations: wear a chemical protective, full-body encapsulating suit and self-contained breathing apparatus (SCBA).

Respiratory Protection: Three types of RPE's are recommended for an H$_2$S environment. All these PPE's are not recommended to be used with air purifying or cartridge type as these can restrict the supply of air.

The only type of Respiratory Protection allowed in an H$_2$S Environment is: Positive Pressure Supplied Air

1. **SCBA (Self Contained Breathing Apparatus):** As is obvious from its name, SCBA are the most famous PPE's used in H$_2$S environment. SCBA is recommended to be used in positive pressure supply. SCBA is rated at 30 minutes. It has 45 cubic feet of air. An employee on by demand will breathe up to 4.5 cubic feet a minute when under stress.

Some of the benefits and limitations of SCBA are:-

▪ These are Portable and have sufficient supply for rescue operations.

▪ These units are heavier due to the oxygen carrying cylinder.

▪ No provisions for loss of air supply. The rescue workers need to evacuate before the air supply finishes.

- SCBA Requires fit test (Almost once per year for each worker who will supposed to be wearing during the contingency). It requires clean shaven and short side burns at respirator seal area.

2. **Air Line Units (SAR):** These units take the air supply from an external source through a hose. These are recommended to be used in H_2S environment as the rescue worker does not have to worry about the shortage of oxygen. An escape pack is mandatory to be used along with SAR work units so that in case the supply of air is disturbed from the external source, the worker may evacuate the area using emergency pack. Some of the advantages and limitations of SAR are as follows:

- These are light weight as no heavy weight cylinder is carried along.

- Large volume of air supply is provided through the hose.

- Escape pack is provided for emergency evacuation in case of failure of SAR.

- Restricted to 300ft of working line (hose). This is because the all PPE's need to be provided with positive pressure i.e. the worker does not need to suck the air resulting extra fatigue on inhalation system. If the length of hose will be increased, the lower will be the positive pressure.

- Long hose may pose tripping hazards for evacuation and rescue workers

- Damaged air hose can cause severe problems for the rescue worker himself.

3. **Escape Packs:** Escape packs are portable units with limited storage of oxygen. These are portable and are used only for evacuation. Some of the advantages and limitations of escape packs are as follows:

- No fit test needed for these packs as these are only used for emergency evacuation

- These escape packs are not intended to be used by rescue workers as they come with min supply of oxygen.

- Low volume of air (5 & 10 minutes).

Respiratory Protection

Up to 100 ppm:

(APF = 10) Any supplied-air respirator*.

(APF = 25) Any powered, air-purifying respirator with cartridge(s) providing protection against hydrogen sulfide.

(APF = 50) Any air-purifying, full face-piece respirator (gas mask) with a chin-style, front- or back-mounted canister providing protection against hydrogen sulfide or Any self-contained breathing apparatus with a full face-piece.

*Reported to cause eye irritation or damage; may require eye protection.

APF = Assigned Protection Factor

Hydrogen sulfide is not released accidentally in lethal quantities very often, but, when there is an unexpected release of a large amount of hydrogen sulfide, it may cause the death of workers.

Although respiratory protective equipment may never be needed, it must be available, and workers must know how to use it in case an emergency occurs which involves the release of hydrogen sulfide.

Intensive training in respiratory protection, in which the worker's physical and psychologic ability to use respirators is confirmed by actual use of the equipment, must be started before the employee begins his assigned work. This training should be repeated as is required (i.e., quarterly) and each time a new crew is formed. All members of a crew ought to receive the same training, even if they have had a previous training session in the same quarter. Repeated practice by those who are already proficient will help to make the drill automatic and ensure swift and accurate reaction in an emergency.

Also, the more experienced workers can serve as examples of proficiency and help to instruct the less experienced workers in the use of respiratory protection. During their 1st year on the Job, workers who are often potentially or actually exposed to hydrogen sulfide in their work (e.g., oil-production or sewer workers) may profit from monthly training and practice wearing and using respirators.

Factor Affecting Performance of PPE's during Escape & Rescue

Following factors may affect the efficiency of PPE's result efficient rescue and evacuation during the emergency.

- **Fit Testing:** The face shield need to be fit tested (once per year) because it requires the Leak tight seal between the face and the mask. No beard or long side burns should be allowed for the workers in the H_2S environment.

- **Training of Use:** H_2S is a silent killer. It does not provide enough time for preparation. It is therefore necessary that workers must be trained how to use these PPE's so that in case of emergency, they should be able to put on these PPE's within minimum possible time to minimize the adverse effects of H_2S.

- **Proficiency Drills:** Proficiency drills must be carried out at periodic intervals so that the workers may manage to put on PPE's in earliest possible time. Only drills and exercises can result into efficient response during the actual emergency situation. Moreover, these drills re-iterate the organization commitment for the safety of its personnel.

- **Maintenance of PPE's**: PPE's must be maintained in serviceable condition.

- **Placement/ Location of PPE's:** PPE's must be placed located in close proximity where all the workers should have easy access. Moreover, one more consideration for the rescue workers is that they should not pass through the H_2S environment while approaching to the PPE's.

Chapter Notes:

CHAPTER 5
First Aid & Rescue Procedures

Escape Procedure

If you are working in an environment where H_2S may be present then you should follow your company specific escape procedure. Some general steps for an escape procedure are as follows:

1. Immediately leave the area and follow your company s escape procedure.

2. Protect yourself by putting an appropriate PPE. You must know the location and use of PPE.

3. Move in cross wind and upwind direction.

4. Proceed to a safe assembly point already communicated to you.

5. Be sure that area is safe from H_2S before you remove your PPE. You may use personal H2S monitor to verify the presence of H_2S or otherwise.

6. Do not return to the work area unless some trained professional allows you after being satisfied that no H2S is present (through portable personal monitor or fixed detection systems). Do not solely rely on your nose which may deceive due to the paralysis of olfactory nerves.

Note: Please do not take part in rescue operation if you are not wearing PPE and you have not been trained for rescue. Do not become another victim.

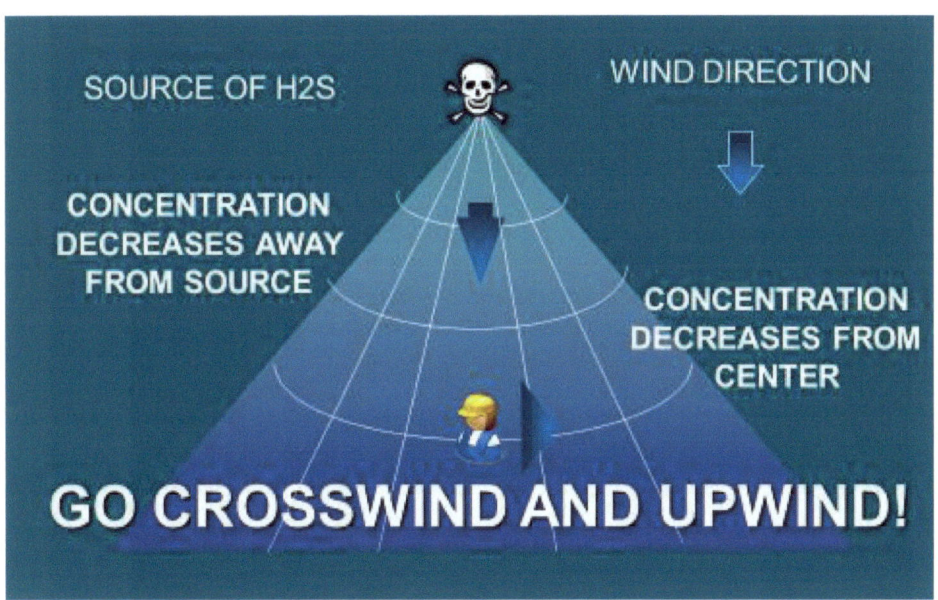

Emergency Situation

If an emergency involving hydrogen sulfide arises, rescuers using respiratory protection shall remove victims to a safe area quickly and initiate appropriate first aid, including artificial respiration if necessary. The victim's lungs should first be cleared of hydrogen sulfide by applying back-pressure artificial respiration briefly before using the more effective mouth-to-mouth artificial respiration. Provision shall be made for prompt transportation to hospital of workers exposed to hydrogen sulfide who have become unconscious, who have respiratory distress, or who feel unwell.

Appropriate local hospitals and medical and paramedical personnel shall be informed by the employer of the possibility of hydrogen sulfide poisoning, even if the chance of emergency is considered remote. Workers sent to the hospital because of hydrogen sulfide exposure shall be identified as such to emergency-room personnel. A qualified medical attendant designated by the employer shall examine all employees who may have been exposed above the occupational exposure limits. Written emergency medical procedures shall be posted where hydrogen sulfide is used.

Rescue Procedure

1. Do not take part in Rescue operations unless you are trained.

2. Wear recommended PPE s before taking part in rescue (Do not become another victim).

3. Provide an appropriate PPE to the victim of to avoid any further exposure.

4. Lift or drag the victim in cross wind upwind direction.

5. If victim has fallen from height then beware as he might have sustained spinal or neck injuries.

6. Ensure the area is safe from H_2S (Portable Monitors) before removing PPEs.

7. If you are not trained for First Aid then seek medical help.

8. Even if an employee has not been "knocked down" by H_2S, it is wise to seek medical attention because of risks of residual problems.

NOTE: These guidelines are for understanding purpose. However, the organization may have their own rescue procedures depending upon their requirement.

First Aid for H₂S victim

Inhalation: Take precautions to prevent a fire (e.g. remove sources of ignition). Also take precautions to ensure your own safety before attempting rescue (e.g. wear appropriate protective equipment). Move victim to fresh air. Keep at rest in a position comfortable for breathing. If breathing is difficult, trained personnel should administer emergency oxygen. DO NOT allow victim to move about unnecessarily. Symptoms of pulmonary edema may be delayed.

If breathing has stopped, trained personnel should begin artificial respiration (AR). If the heart has stopped, trained personnel should start cardiopulmonary resuscitation (CPR) or automated external defibrillation (AED). Avoid mouth-to-mouth contact by using mouth guards or shields. Immediately call a Poison Centre or doctor. Treatment is urgently required. Transport to a hospital.

NOTE: Victims may pose a threat to responders due to the release of hydrogen sulfide from their clothing, skin, and exhaled air.

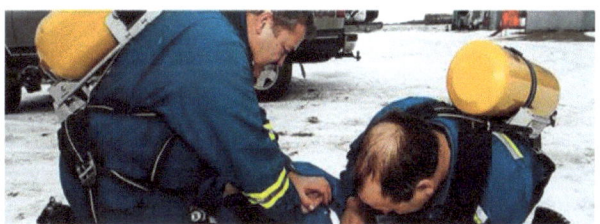

Eye Contact: Immediately flush the contaminated eye(s) with lukewarm, gently flowing water for 15-20 minutes, while holding the eyelid(s) open. DO NOT allow victim to drink alcohol or smoke. Immediately call a Poison Centre or doctor. Treatment is urgently required. Transport to a hospital.

Skin Contact: Liquefied gas: quickly remove victim from source of contamination. DO NOT attempt to rewarm the affected area on site. DO NOT rub area or apply direct heat. Gently remove clothing or jewelry that may restrict circulation. Carefully cut around clothing that sticks to the skin and remove the rest of the garment. Loosely cover the affected area with a sterile dressing. DO NOT allow victim to drink alcohol or smoke. Immediately call a Poison Centre or doctor. Treatment is urgently required. Move the victim to a hospital. Double bag, seal, label and leave contaminated clothing, shoes and leather goods at the scene for safe disposal.

NOTE: Some of the first aid procedures recommended here require advanced first aid training. All first aid procedures should be periodically reviewed by a doctor familiar with the chemical and its conditions of use in the workplace.

Chapter Notes:

CHAPTER 6
Contingency Plan

Preparation for Emergency

- Know the emergency phone numbers and location of safe emergency area.

- Be aware of your contingency program and what part you play in it.

- Remain Wind conscious.

- Know the location of Personal Protective Equipment.

- Know how to read and interpret the H_2S detection system.

- No Smoking policy at the site.

- Be aware of the posted safety signs.

- Always be aware of wind direction and the co-ordinates of the location.

Emergency Response

- Evacuate in an upwind / uphill direction. Report to assembly area immediately.

- Do not return to the area until someone using proper detection equipment has re-evaluated the area and approved it safe to re-enter.

- Only qualified and properly outfitted personnel should attempt rescue.

- If you are not part of the response, evacuate if possible.

Chapter Notes: